**PULBOROUGH
SCHOOL
LIBRARY**

2 0 OCT 1992

Copyright © 1992 by Nord-Süd Verlag AG, Gossau Zürich, Switzerland
First published in Switzerland under the title *Winde wehen, vom Lufthauch bis zum Sturm*
English translation copyright © 1992 by North-South Books

All rights reserved.
No part of this book may be reproduced or utilized in any form
or by any means, electronic or mechanical, including photocopying,
recording or by any information storage and retrieval system
without permission in writing from the publisher.

First published in the United States, Great Britain, Canada,
Australia and New Zealand in 1992 by North-South Books,
an imprint of Nord-Süd Verlag AG, Gossau Zürich, Switzerland.

Distributed in the United States by North-South Books Inc., New York.

Library of Congress Cataloging-in-Publication Data is available
ISBN 1-55858-165-0 (trade binding)
ISBN 1-55858-166-9 (library binding)

British Library Cataloguing in Publication Data
Schmid, Eleonore
The Air Around Us
I. Title
833.914 [J]
ISBN 1-55858-165-0

1 3 5 7 9 10 8 6 4 2
Printed in Belgium

THE AIR AROUND US

WRITTEN AND ILLUSTRATED BY ELEONORE SCHMID

TRANSLATED BY J. ALISON JAMES

NORTH-SOUTH BOOKS / NEW YORK

Nothing is stirring at the edge of the road. No leaf, no bush. Only a butterfly flickers past. The air appears to be still.

But if you look closely, you will see how the grass sometimes shivers when a breath of wind strokes the field.

The sun shines and heats the ground. Where the earth is warm, the air also becomes warm. As it is heated, the warm air rises.

The rising air carries along many water droplets, so tiny that they are invisible. In the sky the drops come together and form clouds.

The air is transparent. You notice it only when it is moving. Then it is called wind.

The wind drives the clouds across the sky. When mountains block the way, the clouds build up. They become heavier and heavier. Soon, the water droplets will get too heavy and will fall back to the earth as rain or snow.

The air high in the mountains is colder than the air in the valley. As the hot air rises into the sky, cool air is pulled down from the mountains. As it moves it becomes a gentle wind.

The wind whispers in the grass. It whistles between stones and howls through caves. It stirs up leaves and sand and makes the trees twist and bend.

Sometimes the wind blows very hard. It rages through the fields and forests. It tackles the trees and tears off their leaves.

The trees creak and moan, and sometimes one crashes to the ground. All the animals hide from the fierce wind. They curl up together and wait for it to pass.

The wind whirls into town. It races over the rooftops and rushes through the grass. It whistles round the corners and howls in the chimneys.

It rattles the doors and clatters the shutters. It tugs at the washing on the line and the clothes are quickly dry.

By the time it reaches the city, the wind calms down. Everyone is outside, enjoying the cool breeze. Children fly kites and try out paper planes.

A hot-air balloon floats slowly across the skyline. The wind brings along the clean mountain air and the smell of the meadows. The sky is a radiant blue.

Cars and trucks roll through the city. They stir up dust and dirt. Their engines blow fumes into the air.

Smoke rises up from factory chimneys. The dirty air hangs over the city. The blue sky becomes grey and the mountains disappear in a haze.

The people in the skyscrapers do not feel the cool breeze from the mountains. Their windows are closed. Heaters and air conditioners make sure that it is never too hot or too cold inside.

The air over the city is filled with planes and helicopters. They fly back and forth, delivering people and packages.

Down by the bay, a thunderstorm has rolled in from the ocean. Dark clouds fill the sky. Lightning sparks, thunder rumbles and a heavy rain begins to fall. The wind rocks the boats and whips up waves on the water.

The rain washes the air and makes it clean again.

Air is as clear as glass. It is everywhere. It covers the entire earth, and it is always moving. Air is very light and yet it has a weight.

High in the mountains the air is so thin that you can hardly breathe. On the moon, and out in space, there is no air.

Every living thing must breathe. People, animals and plants could not live without air. Fresh, clean air makes us healthy; dirty air makes us sick.

Air is priceless. It belongs to everyone. It brings all living things together.

PULBOROUGH
SCHOOL
LIBRARY